International Environmental Labelling

Vol.7

For All DIY & Construction Industries
(Do it yourself " ("DIY") of Building, Modifying, or
Repairing, Renovation, Construction Materials, Cement,
Coarse Aggregates. Clay Bricks, Power Cables,
Pipes and Fittings, Plywood, Tiles, Natural Flooring)

Jahangir Asadi

Vancouver, BC CANADA

Suggest an ecolabel

If you think that we missed a label and/or you are an ecolabelling body, please consider to submit for the next editions of our 11 Volumes International Eco-labelling Book series. Please send your details, and we'll review your suggestions. Our goal is to be as comprehensive as possible, so thank you for your help!

info@TopTenAward.Net

Published by: Top Ten Award International Network
Vancouver, BC **CANADA**
Email: Info@TopTenAward.net
www.TopTenAward.net

Ordering Information:
Quantity sales. Special discounts are available on quantity purchases by universities, schools, corporations, associations, and others. For details, contact the "Sales Department" at the above mentioned email address.

International Environmental Labelling Vol.7/J.Asadi—1st ed.

ISBN 978-1-990451-01-0

Contents

I dedicate this book to my brother in law, Khalil

We hope that, 10,000 years from now, future generations will be able to see flowers that provide bees with nectar and pollen and...
BEES provide flowers with the means to reproduce by spreading pollen from flower to flower...
Jahangir Asadi

Acknowledgements:

I wish to thank my committee members, who were more than generous with their expertise and precious time. I would like to acknowledge and thank the Top Ten Award International Network for allowing me to conduct my research and providing any assistance requested. Special thanks goes to the members of the staff development and human resources department of Bia2Bc Internet Shopping for their continued support.

ABOUT TTAIN

Top Ten Award International Network

Top Ten Award international Network (TTAIN) was established in 2012 to recognize outstanding individuals, groups, companies, organizations representing the best in the public works profession.
TTAIN publishing books related to international Eco-labeling plans to increase public knowledge in purchasing based on the environmental impacts of products.
Top Ten Award International Network provides A to Z book publishing services and distribution to over 39,000 booksellers worldwide, including Apple, Amazon, Barnes & Noble, Indigo, Google Play Books, and many more.
Our services including: editing, design, distribution, marketing
TTAIN Book publishing are in the following categories:

Student
Standard
Business
Professional
Honorary
We focus on quality, environmental & food safety management systems , as well as environmnetal sustain for future kids. TTAIN also provide

complete consulting services for QMS, EMS, FSMS, HACCP and Ecolabeling based on international standards.
ISO 14024 establishes the principles and procedures for developing Type I environmental labelling programmes, including the selection of product categories, product environmental criteria and product function characteristics, and for assessing and demonstrating compliance. ISO 14024 also establishes the certification procedures for awarding the label.

TTAIN has enough experiences to help create new ecolabeling programmes in different countries all over the world.
For more detail visit our website : http://toptenaward.net
and/or send your enquiery to the following email:
info@toptenaward.net

CHAPTER 1

Introduction

This book is dedicated to the subject of environmental labels. The basis for the classification of its parts goes back to the types of environmental labelling according to the classifications provided by the International Organization for Standardization. In each section, while presenting the relevant definitions, I mention the existing international standards and present examples related to each type of labelling. Environmental labelling is an important and significant topic, and its richness is added to every day, which has attracted the attention of many experts and researchers around the world. The idea of compiling this book, came to my mind when I observed that national environmental labelling models have been developed in most countries of the world, but in many other countries, the initial steps have not been taken yet. Therefore, I decided to create the first spark for the development of environmental labelling patterns in other countries by collecting appropriate materials and inserting samples of labelling patterns of different countries of the world. It should be noted that the description of each environmental label in this book does not indicate their approval or denial; they are included only to increase the awareness of all enthusiasts and consumers of the meanings and concepts derived from such labels. We hereby ask all interested parties around the world who wish to start an environmental labelling program in their country to benefit from our intellectual assistance and support in the form of consulting contracts. Increasing human awareness of the urgent need to protect the environment has led to changes in all levels of activities, including the production of marketing products, consumption, use, and sale of goods and services at the national and international levels. Stakeholders involved in environmental protection include consumers, producers, traders, scientific and technological institutes, national authorities, local and international organizations, environmental gatherings, and human society in general. Decisions by consumers and sellers of products are made not only on the basis of key points such as quality, price, and availability of

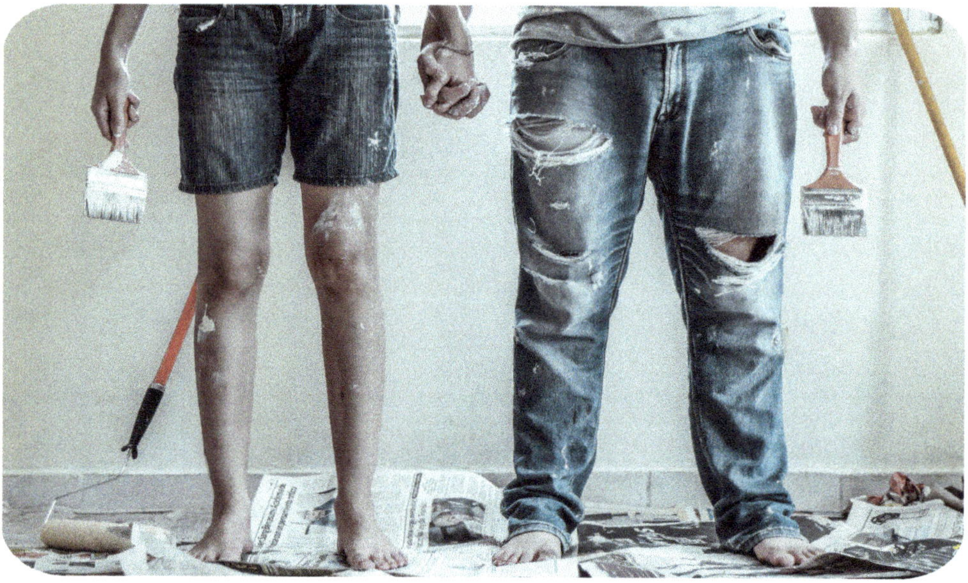

products but also on the environmental consequences of products, including the consequences that a product can have before, after and during production. The most important environmental consequences include water, soil, and air pollution along with waste generation, especially hazardous waste. Further consequences include noise, odor, dust, vibration, and heat dissipation as well as energy consumption using water, land, fuel, wood, and other natural resources. There are further effects on certain parts of the ecosystem and the environment. In addition, the environmental consequences not only include the natural use of the products but also abnormal and even emergency or accidental uses. The basis of studies and studies in this field is done through product life cycle evaluation, which generally involves the study and evaluation of environmental aspects and consequences of a category (product, service, etc.) because of the preparation of raw materials for production until they are used or discarded. Sometimes the phrase "review from cradle to grave" is used for such an evaluation. In addition to the above, the environmental consequences that may occur at any stage of the product life cycle, including the preliminary stages and its preparation, production, distribution, operation, and sale, should also be considered when evaluating it. This type of evaluation refers to product life cycle analysis from an environmental point of view,"

which is a useful tool for measuring the degree of environmental health of a product, comparing different products, improving product quality, and confirming the environmental health claims of the product. The environmental health analysis tool for products and services facilitates their placement in domestic or foreign markets, considering that the awareness of consumers and retailers about the environmental consequences of the product has increased, as has the accurate and explicit measurement by the people in charge at all levels. Local, national, and international in the field of environmental protection. Products that can claim to be environmentally complete in all stages of their life cycle and meet the mandatory and optional environmental needs are considered successful products. Environmental messages refer to the policies, goals, and skills of product manufacturing companies as part of the environmental management systems in which they are applied, and consumers and retailers are increasingly paying attention to this issue when making purchasing decisions. In addition, companies have been encouraged and even forced to adapt their environmental management systems to agencies and retailers and to local, national, international, and other environmental issues.

The environmental health message of a product can be conveyed to the consumer in various ways, including implicitly or explicitly. For example, the implicit or implicit message conveyed directly by the product to the customer is that the product is suitable for the intended use and purpose, and, without material waste in size, weight, and dimensions, is perfectly proportioned and without additional packaging. Sometimes it is necessary to convey these messages and claims about the correctness of the product quite clearly through magazines or other media as well as through certificates that are accurate, simple, and convincing to the consumer in the form of a label. These messages must be accurate and fact-based; otherwise they will nullify the product and create contradictory effects. Confirmation of these claims by a third-party organization will increase its credibility. It should also be noted that the multiplicity of these messages, depending on the type of products or companies producing them, confuses consumers in the market and also creates artificial boundaries or causes a differentiated distinction against certain products or companies. Various models, principles, and methods have been provided by local, regional, national, and international organizations to demonstrate product life cycle analysis and other guidelines on environmental management systems and their labels. At the national level, significant advances have been made in the design of environmental labels in various countries, including developing countries and the Scandinavian countries. For example, the first project was designated in Germany as a Blue Angel in 1977, later on Canada in 1988, the Scandinavian countries and Japan in 1989, the United States and New Zealand in 1990, India, Austria, and Australia in 1991, And in 1992, Singapore, the Republic of Korea, and the Netherlands developed their national environmental labelling. Environmental labels are an environmental management tool that is the subject of a series of ISO 14000 standards. These environmental labels provide information about a product or commodity in terms of its broad environmental characteristics, whether it is about a specific environmental issue or about other characteristics and topics.Interested and pro-environmental buyers can use this information when choosing products or goods. Product makers with these environmental labels hope to influence people's purchasing decisions. If these environmental labels have this effect, the share of the product in question can increase, and other suppliers may create healthy environmental competition by improving the environmental aspects of their products and commodities. The overall goal of environmental labels is to convey acceptable and accurate information that is in no way misleading regarding the environmental aspects of products and commodities, and they encourage the consumer to buy and produce products that reduce stress on the environment. Environmental labelling must follow the general principles that the International Organization for Standardization has published in a collection entitled the ISO 14020 standard, which refers to these general principles here. It should be noted that other documents and laws in this field are considered if they are in accordance with the principles set out in ISO 14020.

Going green in your
home does not have to be a
big budget project. If you're a
hom wner or r r on a bud
get, making little changes to your
life style at home can help you in-
corporate eco-friendly habits

CHAPTER 2

General Principles on Environmental Labelling

1 The First Principle: Evironmental notices and labels must be accurate, verifiable, relevant, and in no way misleading and/or deceptive.

2 The Second Principle: Procedures and requirements for environmental labels will not be ready for selection unless they are implemented by affecting or eliminating unnecessary barriers to international trade.

3 The Third Principle: Environmental notices and labels will be based on scientific analysis that is sufficiently broad and comprehensive, and to support this claim, the product must be reliable and reproducible.

4 The Fourth Principle: The process, methodology, and any criteria required to support the announcements on environmental labels will be available upon request all interested groups.

5 The Fifth Principle: Development and improvement of environmental notices and labels should be considered in all aspects related to the service life of the product.

6 The Sixth Principle: Announcements on environmental labels will not prevent initiative and innovation but will be important in maintaining environmental implementation.

7 The Seventh Principle: Any enforcement request or information requirement related to environmental notices and labels should be limited to the necessary information to establish compliance with an acceptable standard and based on the notification standards and environmental labels.

8 The Eighth Principle: The process of improving the announcement and environmental labels should be done by an open solution with interested groups. Reasonable impressions must be made to reach a consensus through this process.

9 The Ninth Principle: Information on the environmental aspects of the product and goods related to an advertisement and environmental label will be prepared for buyers and interested buyers from a group consisting of an advertisement and an environmental label.

DIY Projects for Your Home

Some people DIY because they love the process.

Some people DIY to save money.

Some people DIY because it's eco-friendly.

And some people, DIY because of all three reasons!

CHAPTER 3

Types of Environmental Labelling

At present, according to the classification provided by the International Organization for Standardization, there are three types of environmental labelling patterns:

1 Type I labelling: This labelling is known as eco-labelling, and because it is difficult to translate this word into many languages, it presents another reason to adhere to a numerical classification system. In the content of Type I labelling, a set of social commitments that creates criteria according to the scientific principles on the basis of which a product is environmentally preferable is discussed. Consumers are then instructed in assessing environmental claims and must decide which packaging is more important.

2 Type II labelling: refers to the claims made on product labels in connection with business centers. This includes familiar claims such as recyclable, ozone-friendly, 60% phosphate-free, and the like. This type of labelling can be in the form of a mark or sentence on the product packaging. Some of them are valid environmental claims—and some can be completely misleading. Usually, all countries have laws against deceptive advertisements, so why has the International Organization for Standardization discussed this issue? The answer is that it is not clear whether the environmental claims have a technical basis or whether the ad is meaningless.

3 Type III labelling: is a distinct form of third-party environmental labelling pattern designed to avoid the difficulties that can result from type-one labelling. Technical committee for Environment of International organization for Standardization has undertaken a new project to standardize guidelines and Type III labelling methods. One of the main objections raised by industries to Type I labelling is the basis for its management.

Eco-Friendly Products Are Cost-Effective

In contrast, green cleaning products are less abrasive and offer cost saving opportunities such as reduced cost on repair and replacement of damaged floors and surfaces, safer work environments, and reduced water and chemical use

CHAPTER 4

Type I Environmental Labelling

Type I labelling: This labelling is known as eco-labelling, and because it is difficult to translate this word into many languages, it presents another reason to adhere to a numerical classification system. In the content of Type I labelling, a set of social commitments that creates criteria according to the scientific principles on the basis of which a product is environmentally preferable is discussed. Consumers are then instructed in assessing environmental claims and must decide which packaging is more important.

Type I adhesive has the following specifications:

A. Has an optional third-party template.

B. When the product meets a certain standard, the labelling of this product is included.

C. The purpose of this program is to identify and promote products that play a pioneering role in terms of environment, which means its criteria are at a higher level than the average environmental performance.

D. Acceptance/rejection criteria are determined for each group of products and are publicly available.

E. The criteria are adjusted after considering the environmental consequences of the product life cycle.

Examples of Type I Labelling:

In this section, and considering the importance of this type of labelling, I provide a description of some examples of Type I labelling related to some countries along with a list of products on which this mark is placed.

Germany

Do you want to live, shop, construct, renovate or equip an office in an environmentally conscious way? The Blue Angel environmental label makes this possible: Around 12,000 environmentally friendly products and services from around 1,600 companies have been awarded the Blue Angel.

If you use products or services holding the Blue Angel ecolabel, you can be sure that you are doing something good for yourself, the environment, and the future. This is because the Blue Angel is an environmental label organised by the federal government of Germany for the protection of people and the environment. It sets very exacting standards, is independent and has proven itself over more than 40 years as a guide for selecting environmentally friendly products.

The Blue Angel is the ecolabel of the federal government of Germany since 1978. The Blue Angel sets high standards for environmentally friendly product design and has proven itself over the past 40 years as a reliable guide for a more sustainable consumption.

„More and more people are mindful of purchasing products that are durable and environmentally friendly, which is precisely what the Blue Angel stands for. For 40 years, the eco-label has guaranteed high standards for the protection of our environment and health - independently and credibly."

Environmentally Friendly Procurement,
Wörlitzer Platz 1, D-06844 Dessau-Roßlau
Tel: +49(0)34021033375
Fax: +49 340 2103 3831
Web:
http://www.blauer-engel.de/en

Ukraine

The ecolabelling program in Ukraine was founded on the initiative of the All-Ukrainian NGO "Living Planet" in 2003. The Green Crane is the first and the only one Type 1 Ecolabel in Ukraine that recognized officially.
The main objective of company's activity is to evaluate the products for compliance with environmental criteria according to ISO 14024 scheme in order to ensure the reliability of data on the environmental benefits of products within a specific category based on the results of the life cycle assessment. Over the 16 years of the program's existence, the Green Crane ecolabel has become a recognizable reliable reference point for consumers and government organizations (in "green procurement" process), as well as effective marketing tool for business.
Program Statistical Information. Today, the program operates with 57 certification standarts in various industries - construction, food, chemical, textile and other. More than 500 certificates have been issued throughout the program's history. Currently, 68 certificates for more than 1,000 products are valid.

Contact:
NGO «Living Planet»
Email: os@ecolabel.org.ua,
info@ecolabel.org.ua
Tel: +380 44 332 84 08
Adress: Magnitogorsky Lane1-B, Kyiv, Ukraine - 02094
Web: https://www.ecolabel.org.ua/en

Finland

EKOenergy is a global, nonprofit ecolabel for renewable energy (electricity, gas, heat and cold). In addition to being renewable, the energy sold with the EKOenergy label fulfils sustainability criteria and helps finance projects that combat energy poverty. The financed projects address several SDGs and give the opportunity for consumers to achieve more with their purchase. The EKOenergy label can be combined with all sourcing methods. EKOenergy-labelled energy is currently available in 40+ countries.

Users of EKOenergy-labelled energy can use the logo on products made with EKOenergy. By using our internationally recognised logo, individuals and companies demonstrate their commitment to renewables.

Contact info:
Steven Vanholme,
steven.vanholme@sll.fi

Catalonia

The Emblem of Guarantee of Environmental Quality identifies the products and services that have passed strict environmental quality criteria that go beyond regulatory requirements and bear in mind the life cycle. This type I ecolabelling system, adapted to ISO 14024, is compatible and on a par with other international ecolabelling systems such as the EU Ecolabel.

The Emblem of Guarantee of Environmental Quality was created in November 1994. Its original scope was guaranteeing the environmental quality of certain product properties and characteristics. In 1998, the scope was expanded to include services.

Through the creation of this ecolabel, Catalonia is eager to lead the way in terms of having its own regional ecolabelling system in Europe in keeping with European countries with a long history in environmental protection.

The purpose of the ecolabel is to encourage the design, production, marketing and consumption of more environmentally friendly products and services.

Contact details
Contact person: Josep M. Masip
josepmaria.masip@gencat.cat
ssq.tes@gencat.cat

Bolivia

Legally established in Bolivia, IMOcert has a presence in more than 20 countries in Latin America and the Caribbean, has regional offices in Peru, Paraguay, Mexico, among others. As an organic control body, it has been accredited for many years in accordance with the NOP / USDA Regulation, it also has accreditation of the ISO / IEC 17065 standard and also has other national accreditations of countries where it operates and authorizations for other sustainable schemes and social. IMOcert has extensive experience in certification of producer groups, actively collaborating in the origin and development of the internal control systems methodology.

Complete contact detail
Nombre/Name: Alberto Levy
Gerente Ejecutivo / Executive manager
Teléfono de oficina/office pone: (+591) 4456880/81
Fax: (+591) 44456882
Correo electrónico/ e-mail: imocert@imocert.bio – alevy@imocert.bio
www.imocert.bio

China

China Environmental United Certification Center (CEC), approved by the Ministry of Ecology and Environment of the People's Republic of China (MEE) and accredited by Certification and Accreditation Administration Committee of PRC, is a comprehensive certification and service institution leading in environmental protection, energy saving and low carbon areas. . CEC is committed to serve building national ecological civilization; and has carried out research on environmental protection, energy saving, low carbon development strategies and solutions; has been continuously improving and innovating green industry evaluation system on industrial green development and transition CEC is building a bridge between green production and green consumption by offering independent, impartial and high-quality evaluation and certification service for government, enterprises and the public. CEC is a state-owned, non-profit, legal entity of independent third-party certification. It integrates the certification resource from the former National Accreditation Center for Environmental Conformity Assessment, the Secretariat of China Environmental Labelling Products Certification Committee, Environmental Development Center of MEE, the Chinese Research Academy of Environmental Sciences and other institutions. Business areas includes: products certification, management systems certification, services certification, addressing climate change, energy-saving and energy efficiency certification, green supply chain assessment, environmental stewardship, green credit assessment and green manufacturing system evaluation. CEC also carries out standard establishment and research project and international cooperation and exchanges, etc.

Contact:
Website: http://en.mepcec.com/
E-mail: zhangxiaoh@mepcec.com , zhangxiaoh@mepcec.com

Philippines

The National Ecolabelling Programme Green Choice Philippines (NELP-GCP) is an ecolabelling programme based on ISO 14024 Guiding Principles and Procedures. It is a voluntary, multiple criteria-based, and third-party programme the aims to encourage clean manufacturing practices and consumption of environmentally preferable products and services. It awards the seal of approval to product or service that meets the environmental criteria established for the product category by a multi-sector Technical Committee. Products with the Green Choice Philippines Seal assures the consumers on its preference for the environment. NELP-GCP is being administered by the Philippine Center for Environmental Protection and Sustainable Development, Inc. (PCEPSDI).

Contact:
Website: https://pcepsdi.org.ph/
E-mail: greenchoicephilippines@pcepsdi.org.ph,
greenchoicephilippines@gmail.com

Sweden

TCO Certified is the world-leading sustainability certification for IT products. Covering 11 product categories including computers, mobile devices, display products, and data center products, its comprehensive criteria are designed to drive social and environmental responsibility throughout the product life cycle. Independent verification of criteria compliance is always included. Independent verifiers spend around 20,000 hours every year on tests and audits. Currently, more than 3,500 products from 27 well-known IT brands are certified. The purpose of TCO Certified is to drive progress toward a future where all IT products have a sustainable life cycle, something that requires a collective effort from IT buyers as well as industry. TCO Certified helps the IT industry structure their work with sustainability and offers a platform for continuous improvement. Organizations that buy IT products use the certification as a tool for making more responsible IT product choices.

Contact:
TCO Development |
Linnégatan 14 | 11447 Stockholm, Sweden |
Mobile: +46 (0) 706 358351|
Email: Marketing@tcodevelopment.com

Taiwan

In order to provide guidance to consumers for the purchase of products with high energy efficiency three principal policies have been employed in the promotion of energy efficiency management for energy-consuming equipment and apparatuses in Taiwan; those include Minimum Energy Performance Standard (MEPS), voluntary energy efficiency labeling program and mandatory energy efficiency rating labeling program. At present, Taiwan has announced MEPS requirements for 29 product categories; and 51 product categories for the voluntary energy efficiency labeling program; 17categories of products for the mandatory Energy Efficiency Rating Labeling system.

The related information is on the website www.energylabel.org.tw.

EKOagros

ORGANIC CERTIFICATION

Lithuania

EKOAGROS is the only institution in Lithuania for more than 20 years carrying out certification and control activities of organic production and products of national quality, also providing services of certification activities in accordance with the foreign national and private standards in foreign countries. From year 2017 EKOAGROS is accredited as certifying agent to conduct certification activities on crops, wild crops, livestock and handling operations in accordance with USDA NOP.

Contact information:
EKOAGROS
Address K. Donelaicio str. 33, LT-44240 Kaunas, Lithuania
Tel. No. +370 37 20 31 81
Website: www.ekoagros.lt

Germany

FSC® is a global not-for-profit organization that sets the standards for responsibly managed forests, both environmentally and socially. When timber leaves an FSC certified forest they ensure companies along the supply chain meet our best practice standards also, so that when a product bears the FSC logo, you can be sure it's been made from responsible sources. In this way, FSC certification helps forests remain thriving environments for generations to come, by helping you make ethical and responsible choices at your local supermarket, bookstore, furniture retailer, and beyond. www.fsc.org

FSC® International
Adenauerallee 134
53113 Bonn
E-mail: info@fsc.org
Phone: +49 (0) 228 367 66

FSC Canada
50 rue Sainte-Catherine Ouest,
bureau 380B, Montreal, QC H2X 3V4
Email: info@ca.fsc.org
Telephone: 514-394-1137

CERTIFIED
Carbonfund.org
A Carbon Neutral Product

USA

The Carbonfree® Product Certification is a meaningful, transparent way for you to provide environmentally-responsible, carbon neutral products to your customers. By determining a product's carbon footprint, reducing it where possible and offsetting remaining emissions through our third-party validated carbon reduction projects, companies can:

- Differentiate their brand and product
- Increase sales and market share
- Improve customer loyalty
- Strengthen corporate social responsibility & environmental goals

The Carbonfree® Product Certification Program is proud to be part of Amazon's Climate Pledge Friendly Program!
Carbonfund.org is leading the fight against climate change, making it easy and affordable to reduce & offset climate impact and hasten the transition to a clean energy future.

Contact:

O: 240.247.0630 ext 633
C: 203.257.7808
M: 853 Main Street, East Aurora, NY, 14052

sustainable florist

Netherland

For more than 25 years, the independent Dutch foundation SMK works from professional knowledge with companies to improve the sustainability of products and business management. SMK cooperates with an extensive stakeholder network of governments, producers, branch and non-governmental organisations, retailers, consultancies, researchers. The SMK Boards of Experts establish objective criteria for more sustainable products and services. SMK's transparent work processes, third party audits and certifications are conducted according to international certification standards, mostly under supervision of the Dutch Accreditation Council. Besides, SMK is Competent Body of the EU Ecolabel. SMK keeps an extensive database of sustainability criteria.

Contact:
Bezuidenhoutseweg 105 - 2594 AC Den Haag
Telefoon: 070-3586300
Mobiel: 06-82311031
(niet op woensdag)
www.smk.nl

Global

Mission:

Our mission is the development, implementation, verification, protection and promotion of the Global Organic Textile Standard (GOTS). This standard stipulates requirements throughout the supply chain for both ecology and labor conditions in textile and apparel manufacturing using organically produced raw materials. Organic production is based on a system of farming that maintains and replenishes soil fertility without the use of toxic, persistent pesticides and fertilizers. In addition, organic production relies on adequate animal husbandry and excludes genetic modification.

The fastest way to learn about GOTS is to watch our four minute Simple Show Clip: https://global-standard.org/resource-library/clips.

Contact detail:

Web: www.global-standard.org
Email: mail@global-standard.org

Denmark, Finland, Norway, Iceland, Sweden

The Nordic Swan Ecolabel

The Nordic Swan Ecolabel is the official Nordic ecolabel supported by all Nordic Governments. It is among the world›s strictest and most recognised environmental certifications.

The Nordic Swan Ecolabel is a Type I environmental labelling program established in 1989 by the Nordic Council of Ministers, connect¬ing policy, people, and businesses with the mission to make it easy to make the environmentally best choice. Nordic Ecolabelling is the non-profit organisation responsible for the Nordic Swan Ecolabel.

The organisation offers independent third-party certification and support for a wide range of product areas and services, ensuring that they comply with the Nordic Swan Ecolabel's strict requirements through documentation and inspections.

30 years of experience and expertise has made the Nordic Swan Ecolabel a powerful tool that paves the way to a sustainable future by giving producers a recipe on how to develop more environmentally sustainable products, and giving consumers credible guidance by helping them identify products that are among the environmentally best.

Globally, you can find more than 25,000 Nordic Swan ecolabelled products. 93% of all Nordic consumers recognise the Nordic Swan Ecolabel as a brand, and 74% believe that the Nordic Swan Ecolabel makes it easier for them to make envi¬ronmentally friendly choices (IPSOS 2019).

Denmark, Finland, Norway, Iceland, Sweden

Securing a sustainable future

The Nordic Swan Ecolabel works to reduce the overall environmental impact from production and consumption and contributes significantly to UN Sustainable Development Goal 12: Responsible consumption and production.

To ensure maximum environmental impact, the Nordic Swan Ecolabel sets product specific requirements and evaluates the environmental impact of a product in all relevant stages of a product lifecycle - from raw materials, production, and use, to waste, re-use and recycling.

Common to all products certified with the Nordic Swan Ecolabel is that they meet strict environmental and health requirements. All requirements must be documented and are verified by Nordic Ecolabelling. Nordic Ecolabelling regularly reviews and tightens the requirements.

Therefore, certifications are time-limited and companies must re-apply to ensure sustainable development.

International website:
Nordic-ecolabel.org
National websites:

Denmark:	ecolabel.dk	
Sweden:	svanen.se	
Norway:	svanemerket.no	(in Norwegian)
Finland:	joutsenmerkki.fi	(in Finnish)
Iceland:	svanurinn.is	(in Icelandic)

Republic of Korea

The Korea Eco-labelling is a certification system enforced by the Ministry of Environment and KEITI(Korea Environmental Industry & Technology Institute). Since its foundation in April 1992, the system has certified a wide range of eco-friendly products, which were selected as excellent not only in terms of their environmental-friendliness, but also for their quality and performance during their life cycle. Korea Eco-labelling is voluntary certification scheme to attach logo to products with superior environmental quality throughout their lifecycle to other products of the same use, and thus to provide product information to consumers. For 30 years, the scheme has launched plenty of eco-labelling product standards covering personal and household goods, construction materials, office equipment furniture, etc. It products categories which cover all aspects of products, such as reduction of use of harmful substances, energy saving, resource saving, etc. As of April 30th 2021, 169 criterias(=standards), and certifications for 18,250 products(4,549 companies) have maintained.

Contact:
Korea Environmental Industry & Technology Institute(KEITI)
Office of Korea Eco-Label Innovation
Address: 215, Jinheung-ro, Eunpyeong-gu, Seoul, Repulic of Korea
T: +82 2 2284 1518
F: +82 2 2284 1526
E: accolly@keiti.re.kr
W: www.keiti.re.kr

EUROPE

Established in 1992 and recognized across Europe and worldwide, the EU Ecolabel is a label of environmental excellence that is awarded to products and services meeting high environmental standards throughout their life-cycle: from raw material extraction, to production, distribution and disposal. The EU Ecolabel promotes the circular economy by encouraging producers to generate less waste and CO2 during the manufacturing process. The EU Ecolabel criteria also encourages companies to develop products that are durable, easy to repair and recycle.
The EU Ecolabel criteria provide exigent guidelines for companies looking to lower their environmental impact and guarantee the efficiency of their environmental actions through third party controls. Furthermore, many companies turn to the EU Ecolabel criteria for guidance on eco-friendly best practices when developing their product lines. The EU Ecolabel helps you identify products and services that have a reduced environmental impact throughout their life cycle, from the extraction of raw material through to production, use and disposal. Recognised throughout Europe, EU Ecolabel is a voluntary label promoting environmental excellence which can be trusted.

Spain , Germany, Italy, Sweden, Greece, Portugal, Poland, Belgium, Netherlands, Estonia, Finland, Austria, Lithuania, Czech Republic, Norway, Cyprus, Ireland, Slovenia, Hungary, Romania, Croatia, Bulgaria, Malta, Slovak Republic, Latvia, Luxembourg, Iceland

Contact and more information via: http://ec.europe.eu

INVOLVES

CHAPTER 5

Type II Environmental Labelling

Type II environmental labelling refers to the claims made on product labels in connection with business centers. This includes familiar claims such as recyclable, ozone-free, 60% phosphate-free, and the like. This type of labelling can be in the form of a mark or sentence on the product packaging. Some of them are valid environmental claims—and some can be completely misleading.

Usually, all countries have laws against deceptive advertisements, so why has the International Organization for Standardization discussed this issue? The answer is that it is not clear whether the environmental claims have a technical basis or whether the ad is meaningless.

Most countries have guidelines at the national level to help producers and consumers know what constitutes a true, scientifically valid claim.
There is a national standard on this in Canada. In Australia, the Consumer Commission has published guidance on this, and there are similar examples in other countries.

USA

The original recycling symbol was designed in 1970 by Gary Anderson, a senior at the University of Southern California as a submission to the International Design Conference as part of a nationwide contest for high school and college students sponsored by the Container Corporation of America. The recycling symbol is in the public domain, and is not a trademark. The Container Corporation of America originally applied for a trademark on the design, but the application was challenged, and the corporation decided to abandon the claim. As such, anyone may use or modify the recycling symbol, royalty-free.

For More information refer to ISO 14021,
Environmental Labels and declarations

Canada

Environmental Sustain for Future kids established in Vancouver, BC Canada in 2020. (ESFK) is an international ecolabel focused on taking care of environment for future of kids.

ESFK defined as 'self-declared' environmental claims made by manufacturers and businesses based on ISO 14020 series of standards, the claimant can declare the environmental objectives and targets in relation to taking care of environment for future kids. However, this declaration will be verifiable.

Environmental Sustain for Future Kids
Vancouver, BC CANADA

Email: info@esfk.org
Web: www.esfk.org

Canada

Energy Moon Ecolabel established in Coquitlam, British Columbia, Canada in 2021. ENERGY MOON is an international ecolabel focused on Energy Saving in different industries and categorized as Type II Environmental Labelling. It's defined as 'self-declared' energy saving claims made by manufacturers and businesses based on ISO 14020 series of standards, the claimant can declare the Energy Saving objectives and targets and also propose programmes for achiving the defined objectives. However, this declaration will be verifiable.

Energy Moon
Coquitlam, BC CANADA

Email: info@energymoon.org
Web: www.energymoon.org

The greatest expense of any kitchen remodel may very well be your cabinets. Instead of replacing them altogether, consider repainting them or simply refacing them. Most likely your cabinets are in great conditions. New doors and drawers can give seemingly tired cabinets a whole new life!

CHAPTER 6

Type III Environmental Labelling

Type III environmental labelling is a distinct form of third-party environmental labelling pattern designed to avoid the difficulties that can result from type I labelling. Technical committee for Environment of International organization for Standardization has undertaken a new project to standardize guidelines and Type III labelling methods. One of the main objections raised by industries to Type I labelling is the basis for its management.

Due to the nature of the system, less than 50% of the various products on the market can meet the criteria and qualify for Type I Labelling. As long as the industry is the main supporter of other third-party models for quality systems, it is sometimes difficult for an industry to support a program that can only benefit 15% of its members. This type of labelling is currently practiced in some countries, such as Sweden, Canada, and the United States. Choosing the right product has never been easy, but Type III labelling will help because each product can have a label that describes its environmental performance and is certified by a third-party company. Consumers can then compare labels and choose their favorite products.

CHAPTER 7

All about 'Eco-friendly' DIY Building, Modifying, and/or Reconstruction

DIY is short for do-it-yourself. It means carrying out home repairs, maintenance, and improvements yourself instead of hiring a professional. Interest in DIY took off after the Second World War. Changes such as growth in home ownership and the arrival of TV programs about home improvement helped to fuel the DIY movement.

After a while our homes need a change or some updating. They can seem tired and old and in need of a little refreshing. Ideally, perhaps, you might like to move to a larger home but the economy has you worried and you want to spend your money wisely. So instead, you've decided to renovate your home not only to perk it up but to better accommodate your current needs and lifestyle. If you haven't thought to do so already, you might want to think about some environmentally savvy ways to renovate your home.

Top Ten Eco-Friendly Ways to Renovate Your Home

If you are planning on renovating your home, you'll want to make sure you do it in style but without further causing harm to our environment. Contrary to what many people think, there are plenty of ways to make your home renovation an environment-friendly one. That way, you have a home that is not only aesthetically pleasing, but one that reduces environmental impact. But how exactly do you renovate a home to make it simultaneously eco-friendly and stylish? we show you Top Ten of eco-friendly ways to renovate your home:

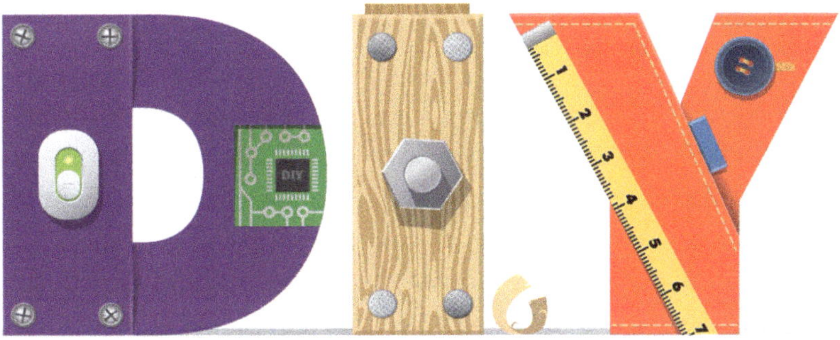

1. Use recycled glass

There are many home depots that now cell bio-glass that look like new windows but are 100% environment-friendly. They make a fantastic addition to the kitchen or your bedroom, as it brightens the whole space allowing natural lighting and morning sunshine – a great way to start your morning!

2. Use formaldehyde-free cabinets

Formaldehyde is commonly used in building materials and household products. But even if it's found in many household and beauty products, it's actually quite toxic! We recommend that you look for those free of this arterial to ensure safety for both your home and the environment's well-being. There are now many stores that offer VOC and formaldehyde-free furniture such as kitchen cabinets.

3. Paint with low-VOC or VOC-free paint

VOC is short for volatile organic compounds which produce harmful molecules. This ends up with you and the household incurring long-term health effects, and they aren't good ones either! Using VOC-free products like paint will help you breathe easier when at home.

4. Go solar

The sun is very powerful and it's a renewable source of energy – so use it to your advantage! Collecting the power of the sun via solar panels can give you electricity to last the whole night and you can use it to heart's content completely guilt-free! Not only will you help the environment through saving energy, but you will notice your electricity bills becoming less expensive as well. (For complete detail in this regard refer to IEL Vol.2 Energy)

5. Deconstruct your home – don't demolish it!

If you plan on tearing down walls or even knocking down entire rooms, walk around your home first to see what you can salvage and re-use beforehand. Not only is this eco-friendly, but it will save money in the end. If it ain't broke, don't fix it! Most likely there is a ton of material you can salvage and re-use. Consider everything from light fixtures, to flooring, tile, bricks, cabinets and molding. If you plan on replacing the chandelier in your dining room, instead of tossing it, think about using it in another room – maybe your kitchen, your daughter's bedroom, even a bathroom!

6. Choose bamboo flooring

What makes bamboo different from other types of wood materials? It's durable, moisture-resistant, grows back faster than wood and growing it uses less pesticides. We have an abundance of bamboo that can be harvested without destroying its roots, making it an environmentally friendly option. You get to save the lives of other trees and old growth forests, all while getting sleek chic flooring. Just make sure to get a good hammer to carry out this project; plus, it's energy-saving as well!

7. Using salvaged wood or discarded metal & Donate your unwanted items

Using recycled materials such as wood and metal will help to reduce waste and the need for fossil fuels as trucks and machinery aren't required to cut down existing trees. Plus, it gives your home a unique and contemporary look. Don't worry about rust or damage, as there are many types of salvaged materials that hold durability and beauty. So you really don't want that dining room chandelier in any other room. So you really don't want that dining room chandelier in any other room. Don't toss it, Perhaps you even have a crafty friend who might enjoy repainting and re-purposing it. With this in mind, not only are you being environmentally friendly, but you are truly giving back to the community.

8. Focus on energy-efficient appliances

If you're planning to replace refrigerators, air conditioners, or other types of major appliances, focus on those that help save both the environment and energy. Investing in energy efficient white goods will go a long way to saving you money in the longer term! (For complete detail in this regard refer to IEL Vol.2 Energy)

9. Consider buying pre-owned materials

Habitat for Humanity is one such retailer, but there are many across the country, some even specialize in high-end products. This can be a great and cost-effective way to redo your home. If, in the end, that SubZero fridge is an absolute must have, you could save thousands of dollars buying one that has been used for a couple of years. Cabinets may be the largest expense of a kitchen renovation, these salvage shops often have high quality cabinets in fabulous condition. It's an idea certainly worth investigating.

10. Re-face instead of replace

As I just mentioned, the greatest expense of any kitchen remodel may very well be your cabinets. Instead of replacing them altogether, consider repainting them or simply refacing them. Most likely your cabinets are in great conditions. New doors and drawers can give seemingly tired cabinets a whole new life!

Conclusion:

Looking for a way to renovate your home while saving the environment isn't all that difficult if you follow these eco-friendly renovation tips. And these helpful hints don't just apply to your home – they can apply to your office, work space or any other building project you're thinking of undertaking!

Do you have any other tips or suggestions on how to renovate your home in a more sustainable way? Please let us know for considering in the next edition of this International Environmental Volume Series.

Eco-house

Most Popular Eco-Friendly Flooring Solutions

We have provided a guide of the most popular eco-flooring solutions, some are new, some are old and a few will make you think:

1. Cork

Cork flooring is a product made from the bark of the cork oak tree, a material which is ground, processed into sheets and baked in a kiln to produce tiles that serve as flooring for offices, light commercial locations, and residences. Cork is harvested from the bark of the cork oak tree commonly found in the forests of the Mediterranean. The trees are not cut down to harvest the bark, which will grow back every three years, making it an ideal renewable source. It has anti-microbial properties that reduce allergens in the home, is fire retardant, easy to maintain and acts as a natural insect repellent too.

2- Bamboo

Bamboo flooring is another wood like option that is gaining in popularity. It is actually a grass that shares similar characteristics as hardwood. It is durable, easy to maintain and is easy to install. Bamboo is sustainable and made from natural vegetation that grows to maturity in three to five years, far less than the twenty years trees can take.

Bamboo, while usually very light, is available in many hues that will work in any setting or decor. Its varied grains and wide array of colors give it an edge over traditional flooring by allowing for customization not often found elsewhere.

FLOOR AND MATERIAL ICONS

3. Linoleum

Linoleum is one of the most natural and sustainable flooring solutions on the market, appreciated for its natural beauty, comfort and durability for over 150 years. When one thinks of linoleum flooring, vinyl tends to come to mind and yet the two are nowhere close to each other. Vinyl is a synthetic made of chlorinated petrochemicals that are harmful. Linoleum is created from a concoction of linseed oil, cork dust, tree resins, wood flour, pigments and ground limestone.

Like cork, it is fire retardant and water resistant. Linoleum is not new to the market; it fell out of favor with the introduction of vinyl in the 1940's. As architects and designers began asking for it again, it reemerged with a vast array of bright vibrant colors and a new sealer to protect it from stains. It has a long shelf life and will hold up to a lot of wear and tear. In addition to being made with renewable materials, linoleum is biodegradable and won't take up space in landfills. Linoleum does not emit harmful VOCs (brand new linoleum does have a harmless odor from the linseed oil content that dissipates after a few weeks).

4. Glass tiles

Ever wonder what happens to the beverage bottles that are shipped to the recycler? They are converted into beautiful glass tiles. This renewable source is fast becoming a wonderful option for floors as well as bathroom and kitchen walls. Glass has similar benefits of other eco-friendly materials. It is non-absorptive and won't mildew or mold in damp environments. It is easy to maintain and won't stain.

Glass comes in a limitless array of colors, patterns and finishes suitable for most design schemes. Unlike ceramic tiles, glass will reflect light rather than absorb it, adding that additional layer of light some rooms need.

5. Concrete

Polished concrete is an unlikely sustainable material that is gaining in popularity. Concrete is typically slab on grade and used as a sub flooring in some residential settings. If it is polished and tinted to the homeowners taste and style there is no need for traditional flooring to be put over it.
From creating a tiled effect with different colors to inlaying other materials such as glass the design possibilities are endless. Concrete is extremely durable, easy to clean and never needs to be replaced.

CONCRETE SLAB ICON SETS.

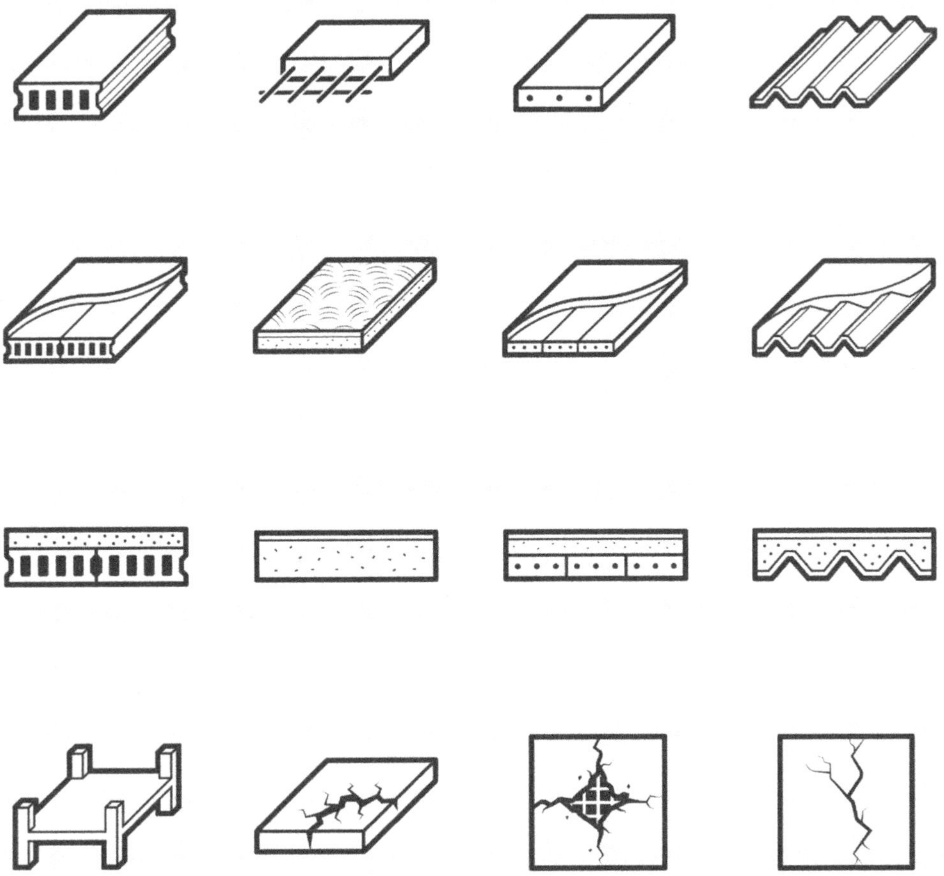

6. Organic wool carpet

Carpet has long been a favorite go-to material for most homes. It is soft to walk on, comfortable to sit on and comes in a range of colors and patterns. Unfortunately, carpet has typically been made using volatile chemical compounds or toxins that are harmful to the environment and to our health. There are eco-friendly options though.

Consider carpets made of organic wool. Organic wool is a natural resource spun into a thread that can be dyed any color imaginable, and then be woven to create a carpet. It is one of the first materials to be used as a floor covering, is very durable and can last centuries. In some families wool rugs have been passed down from generation to generation making them family heirlooms. At the end of its useful life, the pile from wool carpet can be returned to the ground, where the nutrients released as it decomposes promote further grass growth, and the natural production cycle starts all over again. Under the right conditions, wool is totally biodegradable.

7. Corn Carpet

One of the newest environmentally fibers, Carpet is made of corn sugar. It is a very high performing carpet fiber. It is one of the best in terms of durability and the best in terms of stain resistance (in our opinion). Since utilizing corn sugar may take away from the animal and human feed, this may make it not as environmentally friendly as the previous two options. However, you could argue that the durability of Sorona Smartstrand makes it more eco-friendly than PET polyester. It depends on how you look at it. This is a very exciting new carpet fiber.

Corn carpets are made of 35% renewable materials. What are “renewable materials”? It typically refers to a plant because they can be re-grown. In this case, corn sugar is the renewable component. These carpets are made from propanediol, or Bio-PDO, a corn-based polymer. It is made from corn sugar, a by-product of making ethanol, generated at a nearby plant.

An E. coli bacteria --genetically-modified by DuPont scientists--breaks down the corn sugar through a fermentation process that is much like making beer.

DIY
How to make homemade natural wall coating from recycled clothes and fabrics

There are many reasons to use natural wall coating on your walls; like it being odor-free, sound insulation, moisture insulation, heat and cold insulation, crack resistant and quick and easy applicability. This natural coating can cover glass, ceramic, cement, brick, wood. It also has color diversity, acoustic insulation, high resistance against humidity, applicable on any surface, odore free, repairable, 100 % natural, clean environment, thermal Insulation, fire resistance, no more insect, light and many other benefits.

Recipes for homemade natural wall coating:
If you're wanting to ditch those toxic, commercial chemical paints for coating walls and switch to a more natural, homemade wall coating: this simple recipe will have you coating green in no time. However, before we get to the wall coating, let's check out some of the most common (and most useful) non-toxic products:

Small Shredder
Firstly, you need to purchase a $50 DIY shredder to crush recycling (if your paper shredder is strong enough for fabrics and used clothes, you can use it). It will be around $50 only if you have access to a waterjet cutter, but you most likely don't and the minimum will round to about $150. You can have your home shredder for textile, plastic and metal.

1
2
3
4

DIY: Steps for making homemade Eco friendly wall coating with used clothes and fabrics

Necessary items for start wall coating

- Recycled Fabrics and clothes
- Small home Bia2BC Shredder
- Added to Bia2BC powder and required water
- A plastic pan or basin in which shredded clothes, Bia2BC powder and water can be mixed.
- A medium or large size Bia2BC trowel or plastic trowel for rubbing and smoothing Bia2BC paste on the desired surface.

How to prepare Bia2BC dough:

Do the following steps in order and carefully:

Step 1: Collect used fabrics and clothes for as much that is needed

Step 2: Give it to the small home Bia2BC Shredder and start to shredding above mentioned fabrics and clothes

Step 3: Add required amount of Bia2Bc powder

Step 4: Make and mix right amount of water with items mentioned in Step 1 and 2 in the existing plastic pan.

Step 5: After 15-30 minutes the dough is ready to be installed on the desired surface.

Step 6: The correct way to rub Bia2BC on the desired surface is to: first, thoroughly clean the surface of the wall to remove soot, dust or any obstacles to the adhesion of Bia2BC paste. Then, with a special Bia2BC trowel, spread the Bia2BC dough on the surface until it reaches a thickness of about one millimeter.

Step 7: Be careful in doing this and do not give up quickly because you may not be able to do it well the first time, however, with practice and repetition you can achieve a skill that covers both the entire surface with the dough and the thickness of the dough is minimal (not more than 0.5 mm to 1.5 mm).

Note: We've done our absolute best to provide the best information possible, but since we haven't tried every single one of these DIY items in every possible situation, we can't vouch for them 100 percent. So please be cautious and take many safety pre-cautions.

Port Coquitlam, British Columbia, Canada

Consumers are increasingly demanding sustainable products that are not toxic to themselves or the environment. The natural market is growing exponentially, and choosing raw, natural materials will cement your brand as a safe choice — both environmentally and economically

CHAPTER 9

Top Ten Award International Network Environmental Pioneers

Top Ten Award international Network (TTAIN) was established in 2012 to recognize outstanding individuals, groups, companies, organizations representing the best in the public works profession. TTAIN publishing books related to international Eco-labeling plans to increase public knowledge in purchasing based on the environmental impacts of products. We introduce in each volume some of the organizations that are doing their best in relation to taking care of the environmnet.

Bia2BC Internet Shopping

CANADA

An innovative Canadian-based company providing home recycling tools and equipments and also introducing new methods in recycling of different items. Whether it's coffee capsules from your home, pens from a school, or plastic gloves from a manufacturing facility, Bia2BC is trying to help for collecting and recycling almost any form of waste. One of the most important idea of them is providing home small shredders devices with innovated procedures to recycle used clothes and fabrics to wall coatings.

Contact:
Web: http://Bia2BC.ca
Email: info@Bia2BC.ca

Note: We've done our absolute best to provide the best information possible, but since we haven't tried every single one of these DIY items in every possible situation, we can't vouch for them 100 percent. So please be cautious and take many safety pre-cautions.

UN environment programme

UNEP

The United Nations Environment Programme (UNEP) is the leading global environmental authority that sets the global environmental agenda, promotes the coherent implementation of the environmental dimension of sustainable development within the United Nations system, and serves as an authoritative advocate for the global environment.

Our mission is to provide leadership and encourage partnership in caring for the environment by inspiring, informing, and enabling nations and peoples to improve their quality of life without compromising that of future generations.

Headquartered in Nairobi, Kenya, we work through our divisions as well as our regional, liaison and out-posted offices and a growing network of collaborating centres of excellence. We also host several environmental conventions, secretariats and inter-agency coordinating bodies. UN Environment is led by our Executive Director.

We categorize our work into seven broad thematic areas: climate change, disasters and conflicts, ecosystem management, environmental governance, chemicals and waste, resource efficiency, and environment under review. In all of our work, we maintain our overarching commitment to sustainability.

Website: www.unep.org

SERVICE
PAINTING
SERVICE
HOME REMODELING SERVICE
all you need for home remodeling
HAND TOOLS

Bibliography:

Amberg, N.; Magda, R. Environmental Pollution and Sustainability or the Impact of the Environmentally Conscious Measures of International Cosmetic Companies on Purchasing Organic Cosmetics. Visegrad J. Bioecon. Sustain. Dev. 2018, 1, 23.

Asadi, J., "International Environmental Labelling, Economic Consequencies, Export Magazine, July 2001

Asadi, J. 2008. Mobile Phone as management systems tools, ISO Magazine, Vol.8, No.1

Asadi, J., Eco-Labelling Standards, National Standard Magazine, Sep. 2004.

Balló, Z. Review of the competition between diy store review of the competition between diy store chains in Hungary. Delhi Bus. Rev. 2010, 11, 13–28.

Barbieux, D.; Padula, A.D. Paths and Challenges of New Technologies: The Case of Nanotechnology-Based Cosmetics Development in Brazil. Adm. Sci. 2018, 8, 16.

Basketter, D.; Corsini, E. Can We Make Cosmetic Contact Allergy History? Cosmetics 2016, 3, 11.

Bennett, A.; Guerra, P. DIY Cultures and Underground Music Scenes. Abingdon, Oxon; Routledge: New York, NY, USA, 2019; p. 315.

Benitta Christy P & Dr. Kavitha S, "GO-GREEN TEXTILES FOR ENVIRONMENT", Advanced Engineering and Applied Sciences: An International Journal 2014; 4(3): 26-28

Chemical Week, 1999. Europe's Beef Ban Tests Precautionary Principle. (August 11).

Chaudri, S.K.; Jain, N.K. History of Cosmetics. Asian J. Pharm. 2009, 7–9, 164–167.

CHOI, J.P. Brand Extension as Informational Leverage. Review of Eco- nomic Studies, Vol. 65 (1998), pp. 655-669.

Conway, G. 2000. Genetically modified crops: risks and promise.

Corrado, M., (1989), The Greening Consumer in Britain, MORI, London

Corrado, M., (1997), Green Behaviour – Sustainable Trends, Sustainable Lives?, MORI, london, accessed via countries. Manila, Asian Development Bank 33p.

Cosmetics, Perfume, & Hygiene in Ancient Egypt. Available online: https://www.ancient.eu/article/1061/cosmetics-perfume--hygiene-in-ancient-egypt/ (accessed on 4 May 2017).

Davies, Clive. Chief, Design for the Environment Program, Environmental Protection Agency. Interview. March 24, 2009.

Emilien, G.; Weitkunat, R.; Lüdicke, F. Consumer Perception of Product Risks and Benefits; Springer:

Berlin/Heidelberg, Germany, 2017; p. 596.
Fox, R.W.; Lears Jackson, T.J. The Culture of Consumption: Critical Essays in American History, 1880–1980; Pantheon Books: New York, NY, USA, 1983; p. 236.
Federal Trade Commission, "Sorting Out Green Advertising Claims." http://www.ftc.gov/bcp/edu/pubs/consumer/general/gen02.shtm (March 26, 2009, March 27, 2009)
MSNBC, "Do You Know What's in Your Cleaning Products?" http://today.msnbc.msn.com/id/29663739/ (March 17, 2009)
Ooyen, Carla. Research Manager with Nutrition Business Journal. Personal correspondence. March 19, 2009.
Tekin, Jenn. Marketing Manager with Packaged Facts & SBI. Personal correspondence. March 17, 2009.
University of California - Berkeley. http://berkeley.edu/news/media/releases/2006/05/22_householdchemicals.shtml (March 26, 2009)
U.S. Department of Health and Human Services, Household Products Database.http://householdproducts.nlm.nih.gov/cgi-bin/household/prodtree?prodcat=Inside+the+Home (March 17,
Women's Voices of the Earth, "Household Cleaning Products and Effects on Human Health."http://www.womenandenvironment.org/campaignsandprograms/SafeCleaning/safecleaninghealth (March 17, 2009)
EMONS, W. Credence Goods Monopolists. International Journal of In- dustrial Organization, Vol. 19 (2001), pp. 375-389.
European Union official website: https://ec.europa.eu/info/about-european-commission/contact_en
Feenstra, R.C. "Exact Hedonic Price Indexes," Review of Economics and Statistics 77 (1995): 634-653.
Feenstra, R.C., and J.A. Levinsohn. "Estimating Markups and Market Conduct with Multidimensional Product Attributes," Review of Economic Studies (62 (1995): 19-52.
Forest Stewardship Council: "Principles and criteria for forest stewardship" Document 1.2: <http://www.fscoax.org>
Forsyth, K. 1999. Will consumers pay more for certified wood products? Journal of Forestry 97 (2) : 18-22.
Freeman, A. M III. The Measurement of Environmental and Resource Values. Theory and Methods. Washington D.C.: Resource for the Future, 1993.

Friends of the Earth, 1993. Timber certification and eco-labeling. London, FOE:

Geetha Margret Soundri, "Ecofriendly Antimicrobial Finishing of Textiles Using Natural Extract", Journal of International Academic Research For Multidisciplinary, ISSN: 2320 – 5083, 2014, Vol 2.

Graves, P., J.C. Murdoch, M.A. Thayer, and D. Waldman. "The Robustness of Hedonic Price Estimation: Urban Air Quality," Land Economics 64(1988): 220-233.

Halvorsen, R. and R. Palmquist. "The Interpretation of Dummy Variables in Semilogarithmic Equations." American Economic Review 70:474-75 (1980).

Imhoff, Dan, and Grose, Lynda, and Carra, Roberto., "Organic Cotton Exhibit," Mimeo. Simple Life and distributed the Texas Organic Cotton Marketing Cooperative, O'Donnell, Texas (1996).

Imhoff, Dan. "Growing Pains: Organic Cotton Tests the Fibre of Growers and Manufacturers Alike," reprinted on Simple Life's web page (simplelife.com), but first printed by Farmer to Farmer, December 1995.

Incomplete Consumer Information in Laboratory Markets. Journal of Environmental labeling.

ISO 14020, ISO 14021,ISO 14024,ISO 14025, International Organization for Standardization.

Kennedy, P.E. "Estimation with Correctly Interpreted Dummy Variables in Semilogarithmic Equations," American Economic Review 71: 801 (1981).

Kirchho®, S., (2000), Green Business and Blue Angels.

Kraus, Jeff. Lab Technician at the North Carolina School of Textiles.

Labeling Issues, Policies and Practices Worldwide.

Lamport, L. 1998. The cast of (timber) certifiers: who are they? International J. Ecoforestry 11(4): 118-122.

Large Scale impoverishment of Amazonian forests by logging and fire. 1999.

Lathrop, K.W. and Centner, T.J. 1998. Eco-labeling and ISO 14000: An analysis of US regulatory systems and issues concerning adoption of type II standards. Environmental

Lee, J. et al. 1996. Trade related environmental measures; sizing and comparing impacts.

Lehtonen, Markku. 1997. Criteria in Environmental Labeling: A comparative Analysis on Environmental Criteria in Selected Labeling Schemes. Geneva, UNEP. 148p.

LIEBI, T. Trusting Labels: A Matter of Numbers? Working Paper Uni versity of Bern, No. 0201 (2002).

Lindstrom, T. 1999. Forest Certification: The View from Europe's NIPFs. Journal of Forestry 97(3): 25-31. London

Losey, J.E., Rayor, L.S. & Carter, M.E. 1999. Transgenic pollen harms monarch larvae. Nature 399 20 May): p.214.

Management 22 (2) : 163-172.
Mattoo, A. and H. V. Singh, (1994), Eco-Labelling: Policy Considera-Michaels, R. G., and V. K. Smith. “Market Segmentation And Valuing Amenities With Hedonic Models: The Case Of Hazardous Waste Sites,” Journal of Urban Economics, 1990 28(2), 223-242.
Nicholson-Lord, D., (1993) 'Tis the Season to be Green, The Independent, 20 December
Nuttall, N., (1993), Shoppers can cross green products off their lists, The Times, 3 July
OCDE/GD(97)105. Paris, OECD. 81p.
OECD. “Ec-labelling: Actual Effects of Selected Programmes,” OCDE/GD (97) 105, 1997, Paris. (available on line at http://www.oecd.org/env/eco/books.htm#trademono)
OECD. 1997a. Case study on eco-labeling schemes. Paris, OECD (30 Dec):
OECD. 1997b. Eco-labeling: Actual Effects of Selected Programs.
Osborne, L. “Market Structure, Hedonic Models, and the Valuation of Environmental Amenities.” Unpublished Ph.D. dissertation. North Carolina State University, 1995.
Osborne, L., and V. K. Smith. “Environmental Amenities, Product Differentiation, and market Power,” Mimeo, 1997.
Ozanne, L.K. and Vlosky, R.P. 1996. Wood products environmental certification: the United States perspective”. Forestry Chronicle 72 (2) : 157-165.
Palmquist, R. B., F. M. Roka, and T.Vukina. “Hog Operations, Environmental Effects, and Residential Property Values,” Land Economics 73(1), (1997): 114-24.
Palmquist, R.B. “Hedonic Methods,” in J.B Braden and C.D. Kolstad, eds. Measuring the Demand for Environmental Improvement. Amsterdam, NL: Elsevier, 1991.
Pento, T. 1997. Implementation of Public Green Procurement Programs (22-31) in Greener Purchasing: Opportunities and Innovations. Sheffield, Greenleaf Publ. 325 p.
Perloff, J. “Industrial Organization Lecture Notes,” Mimeo. University of California at Berkeley (1985).
Plant, C. and Plant, J. 1991. Green business: hope or hoax? Philadelphia, New Society Publishers 136 p.
Polak, J. and Bergholm, K. 1997. Eco-labeling and trade: a cooperative approach (Jan.): Policy in a Green Market. Environmental and Resource Economics 22, 419-
Poore, M.E.D. et al. 1989. No timber without trees. London, Earthscan. 352p.
Raff, D. M.G., and M. Trajtenberg. “Quality-Adjusted Prices for the American Automobile Industry: 1906-1940.” NBER Working Paper Series, Working Paper No. 5035, February 1995.
Roberts, J. T. 1998. Emerging global environment standards: prospects and perils. Journal of Developing Societies 14 (1): 144-163.

Rosen, S., "Hedonic Prices and Implicit Markets: Product Differentiation in Pure Competition." Journal of Political Economy. 82: 34-55 (1974).

Ross, B. 1997. Eco-friendly procurement training course for UN HCR. : 126 p.

Ryan, S., and Skipworth, M., (1993), Consumers turn their backs on green revolution, The Times, 4 April

Salzman, J. 1997. Informing the Green Consumer: The Debate over the Use and Abuse of Environmental Labels. Journal of Industrial Ecology 1 (2): 11-22.

Sanders, W. 1997. Environmentally Preferable Purchasing: The US Experience (946-960) in Greener Purchasing: Opportunities and Innovations. Sheffield, Greenleaf Publ. 325p.

Sayre, D. 1996. Inside ISO 14000: The competitive advantage of environmental management. Delray Beach FL., St. Lucie Press. 232p.

SHAPIRO, C. Premiums for High Quality Products as Returns to Reputa- tion. Quarterly Journal of Economics, Vol. 98, No. 4 (1983), pp. 659-680.

Stillwell, M. and van Dyke, B. 1999. An activists handbook on genetically modified organisms and the WTO. Washington DC., The Consumer's Choice Council: 20 p.

Semenzato, A.; Costantini, A.; Meloni, M.; Maramaldi, G.; Meneghin, M.; Baratto, G. Formulating O/W Emulsions with Plant-Based Actives: A Stability Challenge for an Eective Product. Cosmetics 2018, 5, 59.

Teisl, M. F., B. Roe, and R. L. Hicks. "Can Eco-labels tune a market? Evidence from dolphin-safe labeling," Presented paper at the 1997 American Agricultural Economics Association Meetings, Toronto.

THE GERSEN, C. Psychological Determinants of Paying Attention to Eco- Labels in Purchase Decisions: Model Development and Multinational Vali- dation. Journal of Consumer Policy, Vol. 23, No. 4 (2000), pp. 285-313.

Tibor, T. and Feldman, I. 1995. ISO 14000: a guide to the new environmental management standards. Burr Ridge Ill., Irwin Professional Publ. 250 p.

Torre, I. de la, & Batker, D. K. (n.d.) 1999-2000. Prawn to trade: prawn to consume. Graham WA., Industrial Shrimp Action Network (isatorre@seanet.com), [and] Asia –Pacific

Townsend, M. 1998. Making things greener: motivations and influences in the greening of manufacturing. Aldershot, England, Ashgate Publisher. 203p.

U.S. Energy Information Administration, What is U.S. Electricity Generation by Energy Source?, Retrieved From: https://www.eia.gov/tools/faqs/faq.php?id=427&t=3

U.S. Energy Information Administration, Biomass Explained, Retrieved From: https://www.eia.gov/energyexplained/?page=biomass_home

U.S. Environmental Protection Agency. National Water Quality Fact Inventory: 1990 Report to Congress. EPA 503-9-92-006, Apr. 1992.

UK Eco-labelling Board website, accessed via http://www.ecosite.co.uk/Ecolabel-UK/

US Environmental Protection Agency (EPA742-R-99-001): 40 p. <www.epa.gov/opptintr/epp>

US EPA, 1993. Determinants of effectiveness for environmental certification and labeling programs. Washington, D.C., US Environmental Protect

US EPA, 1993. Status report on the use of environmental labels worldwide. Washington, D.C., US Environmental Protection Agency (742-R-93-001 September).

US EPA, 1993. The use of life-cycle assessment in environmental labeling. Washington, D.C., US Environmental Protection Agency (742-R-93-003 September).

US EPA, 1998. Environmental labeling: issues, policies, and practices worldwide. Washington DC., Environmental Protection Agency, Pollution Prevention Division Prepared by Abt

US EPA, 1999. Comprehensive procurement guidelines (CPG) program. Washington, D.C., US Environmental Protection Agency: <www.epa.gov/cpg>

US EPA, 1999. Environmentally preferable purchasing program: Private sector pioneers: How companies are incorporating environmentally preferable purchases. Washington, D.C.,

USG, 1993. Federal acquisition, recycling, and waste prevention. Washington DC., Executive Order: (20 October).

USG, 1998. Greening the government through waste prevention, recycling, and federal acquisition. Washington, D.C., Executive Order 13101 (September).

Kijjoa, A.; Sawangwong, P. Drugs and Cosmetics from the Sea. Mar. Drugs 2004, 2, 73–82. [CrossRef]

Wang, J.; Pan, L.; Wu, S.; Lu, L.; Xu, Y.; Zhu, Y.; Guo, M.; Zhuang, S. Recent Advances on Endocrine Disrupting Eects of UV Filters. Int. J. Environ. Res. Public Health 2016, 13, 782.

Bilal, A.I.; Tilahun, Z.; Shimels, T.; Gelan, Y.B.; Osman, E.D. Cosmetics Utilization Practice in Jigjiga Town, Eastern Ethiopia: A Community Based Cross-Sectional Study. Cosmetics 2016, 3, 40.

Ting, C.T.; Hsieh, C.M.; Chang, H.-P.; Chen, H.-S. Environmental Consciousness and Green Customer Behavior: The Moderating Roles of Incentive Mechanisms. Sustainability 2019, 11, 819.

Chen, K.; Deng, T. Research on the Green Purchase Intentions from the Perspective of Product Knowledge. Sustainability 2016, 8, 943.

Wang, H.; Ma, B.; Bai, R. How Does Green Product Knowledge Eectively Promote Green Purchase Intention? Sustainability 2019, 11, 1193.
Wolf, M.; McQuitty, S. Understanding the Do-it-Yourself Consumer: DIY Motivations and Outcomes. Acad. Mark. Sci. Rev. 2013, 1, 154–170.
Mintel. DIY Review 2005; Mintel: London, UK, 2005.
Miller, D. Material Culture and Mass Consumption; Blackwell: Oxford, UK, 1987; p. 252.
Nguyen, T.T.H.; Yang, Z.; Nguyen, N.; Johnson, L.W.; Cao, T.K. Greenwash and Green Purchase Intention: The Mediating Role of Green Skepticism. Sustainability 2019, 11, 2653.
Cinelli, P.; Coltelli, M.B.; Signori, F.; Morganti, P.; Lazzeri, A. Cosmetic Packaging to Save the Environment: Future Perspectives. Cosmetics 2019, 6, 26.
Eixarch, H.; Wyness, L.; Siband, M. The Regulation of Personalized Cosmetics in the EU. Cosmetics 2019, 6, 29.

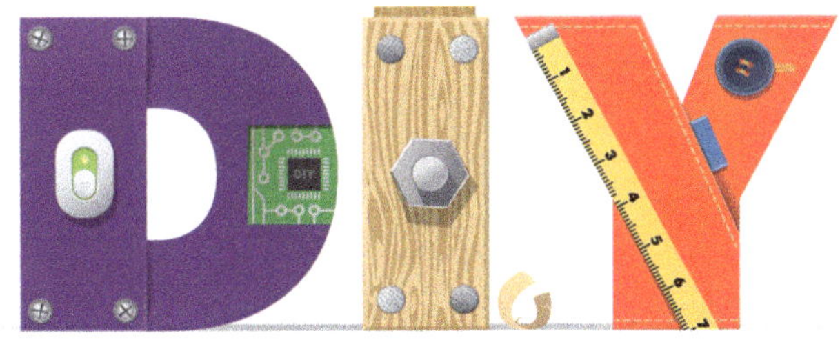

Appendix I: Search by Logos

Here you can search the logos in this volume. It will help you to better undersand the Ecolabels you may encounter while shopping. Buying Eco-products will aid in having a better environment with minimum polution during production processes. Three important parameteres for shopping are **quality**, **price** & **environmental impacts** of the products.

Vol.7 Goto page: 29	Vol.7 Goto page: 46
Vol.7 Goto page: 50	Vol.7 Goto page: 39
Vol.7 Goto page: 35	Vol.7 Goto page: 36
Vol.7 Goto page: 37	Vol.7 Goto page: 40

Vol.7 Goto page: 39	Vol.7 Goto page: 45
Vol.7 Goto page: 50	Vol.7 Goto page: 32
Vol.7 Goto page: 40	Vol.7 Goto page: 34
Vol.7 Goto page: 43 ,44	Vol.7 Goto page: 38

IMOcert	GLOBAL ORGANIC TEXTILE STANDARD · GOTS ·
Vol.7 Goto page: 33	Vol.7 Goto page: 42
sustainable florist	
Vol.7 Goto page: 41	Vol.7 Goto page: 49
ENERGY MOON	
Vol.7 Goto page: 51	Vol.7 Goto page: 31
Vol.7 Goto page: 32	Vol.7 Goto page: 50

APPENDIX II

DIY Algae Bioreactor from Recycled Water Bottles

What is a algae bioreactor used for?
An algae bioreactor is used for cultivating micro or macro algae. Algae may be cultivated for the purposes of biomass production (as in a seaweed cultivator), wastewater treatment, CO2 fixation, or aquarium/pond filtration in the form of an algae scrubber.
DIY Algae Bioreactor from Recycled Water Bottles
Step 1: Make Carbon Dioxide Delivery System. ...
Step 2: Attach Tubing to Manifold. ...
Step 3: Mount Carbon Dioxide System. ...
Step 4: Mount Water Bottles. ...
Step 5: Make Algae Media. ...
Step 6: Media Inoculation. ...
Step 7: Growth and Harvesting.

Seaweed is a large variety of algae that grows in both fresh and salt water. Some seaweeds are edible and can be used in plant fertilizer or medicines. It is possible to grow your own seaweed at home in a large aquarium using salt water you make on the stove.

Appendix III

Environmental Friendly Photos

Environmental friendly photos will be placed in this appendix. These photos can be received in the Top Ten Award International Network inbox from anywhere and everywhere, all over the globe. You can send your appropriate photos to us for them to be considered for publishing in one of the future, related volumes. They will be published with proper credit to the sender. The pictures can also be images of the Ecolabels existing in products within your country.

Eco-house

Vol.1

For All Food Industries
(Meat, Beverage, Dairy, Bakeries, Tortilla, Grain and Oilseed, Fruit and Vegetable, Seafood, And Sugar and Confectionery)

Vol.2

For All Energy & Electrical Industries
(Renewable Energy, Biofuels, Solar Heating & Cooling, Hydroelectric Power, Solar Power, Wind Power, Energy Conservation, Geothermal and Nuclear Power)

Vol.3

For All Fashion & Textile Industries
(Fashion Design, The Fashion System, Fashion Retailing, Marketing and Marchandizing, Textile Design and Production, Clothing and Textile Recycling)

Vol.4

For All Health & Beauty Industries
(Fragrances, Makeup, Cosmetics, Personal Care, Sunscreen, Toothpaste, Bathing, Nailcare & Shaving, Skin Care, Foot Care, Hair Care and Other Health & Beauty Products)

	**Vol.5** For All Maintenance & Cleaning Products (All-purpose Cleaners, Abrasive Cleaners, Powders. Liquids, Specialty Cleaners, Kitchen, Bathroom, Glass and Metal Cleaners, Bleaches, Disinfectants and Disinfectant Cleaners)
	**Vol.6** For All Wood & Stationery Industries (Wooden Products, Cardboard, Papers, Markers, Pens, NoteBooks. Writing Pads and Writing Sets, Pencils, White Papers, Envelopes and Organizers, Staplers and Paper Clips)
	**Vol.7** For All DIY & Construction Industries (Do it yourself ” (“DIY”) of Building, Modifying, or Repairing, Renovation, Construction Materials, Cement, Coarse Aggregates. Clay Bricks, Power Cables, Pipes and Fittings, Plywood, Tiles, Natural Flooring)
	**Vol.8** For All Agricuture & Gardening Industries (Shifting Cultivation, Nomadic Herding, Livestock Ranching, Commercial Plantations, Mixed Farming, Horticulture, Butterfly Gardens, Container Gardening, Demonstration Gardens, Organic Gardening)

Vol.9

For All Professional Products & Services (Teachers, Pilots, Lawyers, Advertising Professionals, Architects, Accountants, Engineers, Consultants, Human Resources Specialist, R&D, Psychologists, Pharmacist, Commercial Banker, Research Analyst)

Vol.10

For All Financial Products & Services (Banking, Professional Advisory, Wealth Management, Mutual Funds, Insurance, Stock Market, Treasury/Debt Instruments, Tax/Audit Consulting, Capital Restructuring, Portfolio Management)

Vol.11

For All Tourism Industries (Airline Industry, Travel Agent, Car Rental, Water Transport, Coach Services, Railway, Spacecraft, Hotels, Shared Accommodation, Camping, Bed & Breakfast, Cruises, Tour Operators)

INTERNATIONAL POCKET GUIDE for HACCP

For All Food Industries (Employees and Employers)

www.ingramcontent.com/pod-product-compliance
Ingram Content Group UK Ltd.
Pitfield, Milton Keynes, MK11 3LW, UK
UKHW020031121026
15625UKWH00042B/292

9 781990 451010